Sensible Steps to Healthier School Environments

Cost-effective, affordable
measures to protect the
health of students and staff.

Contents

Kids Learn Best in Healthy Environments

Approximately 53 million children and 6 million adults in the U.S. spend a significant portion of their days in more than 120,000 public and private school buildings. Many of these buildings are old, in poor condition, and may contain environmental conditions that pose increased risks to the health of children and staff. Reducing exposures to environmental hazards in schools can help children's health. Healthier school environments enable children to learn and produce more in the classroom which can improve their performance and achievements later in life.

This brochure is designed to address some of the most common areas of environmental health concerns found in schools. It also provides one-stop access to learn some facts about these issues and the many existing low cost or no cost, affordable measures, programs and resources available to help prevent, reduce and resolve each of the highlighted environmental hazards. By completing the voluntary Quick Assessment activity provided near the end of this brochure, schools can determine which areas and programs will require more detailed attention. Additionally, by implementing the waste reduction and energy efficiency actions highlighted, schools can conserve valuable, financial resources.

Energy efficiency is a powerful tool that can drastically cut short- and long-term operating costs. At least a ten percent energy savings can occur by implementing little to no-cost minimal actions and energy management practices. School districts can often leverage the opportunity created by energy efficiency upgrades to put in place building upgrades and practices that enhance the health and quality of students' learning environments. Some examples would include improved ventilation systems, moisture control, integrated pest management practices, and removal of PCB-containing lighting ballasts and building materials from school facilities.

Another valuable cost savings tool for schools is waste reduction. Reusing or recycling materials can save schools money in the short term and also encourage environmentally conscious behavior among America's youth. Simple tasks like composting food or yard waste and reusing school supplies can help conserve valuable funds.

Healthy School Environments

EPA's healthy school environments website is designed to provide a one-stop access to the many programs and resources available to help prevent and resolve environmental issues in schools. To learn more about these programs and resources, go to **www.epa.gov/schools.**

Asbestos

Asbestos is the name given to a number of naturally occurring fibrous minerals with high tensile strength, the ability to be woven, and resistance to high heat and most chemicals. Because of these properties, asbestos has been used in a wide range of manufactured goods, including roofing shingles, ceiling and floor tiles, paper and cement products, and textiles. Intact and undisturbed asbestos-containing materials generally do not pose a health risk. Materials containing asbestos may become hazardous and pose increased risk if they are damaged, are disturbed in some manner, or deteriorate over time and release asbestos fibers into building air. Exposure to asbestos is known to cause asbestosis, lung cancer and mesothelioma. Other cancers, primarily of the digestive tract are also possible.

EPA's asbestos program for schools, mandated by the Asbestos Hazard Emergency Response Act (AHERA), and its regulations for schools and other buildings is founded on the principle of "in-place" management of asbestos-containing material (ACM). This approach is designed to prevent asbestos exposure by teaching people to recognize ACM and actively monitor and, where necessary, manage them without removal. Removal of ACM is not usually necessary unless the material is severely damaged or will be disturbed by a building demolition or renovation project.

AHERA requires local education agencies to inspect their schools for asbestos-containing building material and prepare management plans to prevent or reduce asbestos hazards. Public school districts and non-profit private schools (collectively called local education agencies) are subject to AHERA's requirements.

Steps to Reduce Exposure to Asbestos:

✔ Make the school management plan available to all interested parties so they can learn where all identified ACM is located and how it is being monitored.

✔ Ensure all building operations and maintenance staff review the management plan to better understand how to minimize potential disturbance to ACM.

✔ To prevent exposures to asbestos, do not cut, scrape, gouge, drill or physically disturb ACM in any way. Additionally, do not sand grind, saw or abrade ACM in any way.

✔ Report any concerns about damage or deterioration of ACM immediately to the building administrator.

Learn more at:

www.epa.gov/asbestos/pubs/asbestos_in_schools.html

Asthma and Asthma Triggers

Asthma is a disease that affects the lungs and makes it hard for people to breathe. Asthma is a chronic condition and a leading cause of school absenteeism, accounting for more than 10.5 million missed school days per year. On average, one out of every ten school-age children has asthma.

Asthma attacks in schools can be triggered by animal allergens, pest allergens, mold and moisture, dust mites, chemical odors, and, outdoor air pollutants like ozone and particle pollution, or school bus diesel exhaust.

Clutter in classrooms harbors dust. Fabric covered objects such as stuffed animals and pillows are breeding grounds for dust mites. Both dust and dust mites can exacerbate asthma.

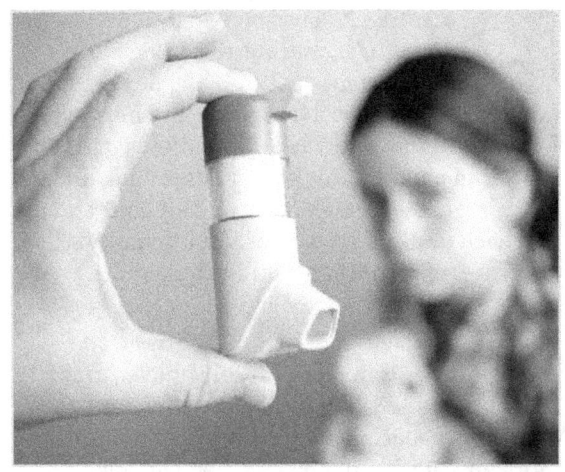

Steps a School Can Take to Reduce Exposures to Asthma Triggers:

✔ Avoid having birds or furry animals such as mice, rabbits or guinea pigs as classroom pets. Fish would make good classroom pets.

✔ Switch to using environmentally friendly cleaning chemicals as much as possible. These products are less likely to have harsh chemical odors that can exacerbate asthma symptoms. Further, always use "wet" dusting techniques wherever possible when cleaning.

✔ Keep classrooms adequately ventilated, free of clutter, dust regularly and frequently wash items that attract dust.

✔ Ensure that your school has an Integrated Pest Management program that will reduce exposures to pesticides while reducing asthma triggers.

Learn more at:

www.epa.gov/asthma/index.html

Buses and Vehicle Idling

Recent EPA emission standards will result in significant reductions in vehicle emissions over the next 15 years. School buses are the safest way for children to get to school. Twenty-five million American children ride school buses daily and on average, these students spend an hour and a half each day in a school bus. Additionally, school buses travel more than 4 billion miles each year.

Buses:

Air pollution from older diesel vehicles and school buses has health implications for everyone, especially children. Children are more susceptible to air pollution because their respiratory systems are still developing and they have a faster breathing rate. In addition to producing a number of hazardous pollutants, diesel exhaust contains significant levels of particulate matter that can deposit into the lungs and can cause lung damage and aggravate respiratory conditions such as asthma.

Vehicle Idling:

Idling vehicles contribute to air pollution and emit air toxins, which are pollutants known or suspected to cause cancer or other serious health effects. This is yet another important issue that affects children's health at school when parents idle their vehicles during student drop-off and pick-up. Exhaust produced by idling vehicles can be pulled into a school through the air intakes of the building's heating, ventilating and air conditioning (HVAC) system where it can accumulate and cause serious health issues for staff and students.

In addition to other environmental benefits, reducing vehicle idling has a number of financial benefits: reduced fuel costs, energy costs and unnecessary engine wear.

Steps to Reduce Vehicle Exhaust at Schools:

✔ Encourage policies to eliminate unnecessary school bus idling.

✔ Upgrade or "retrofit" buses and replace older vehicles with newer, more efficient models (please see http://www.epa.gov/cleanschoolbus).

✔ Establish anti-idling zones for all vehicles at the school (school buses, delivery trucks and parents).

✔ Locate passenger pickup and drop off areas away from a school's air intake supply and classroom windows.

Learn more at:

www.epa.gov/iaq/schools/tfs/guidei.html

www.epa.gov/region8/air/idlefreeschools.html

Carbon Monoxide

Carbon monoxide (CO) is a colorless, odorless gas. It results from incomplete oxidation of carbon in combustion processes. Common sources of CO in schools are improperly vented furnaces, malfunctioning gas ranges, and exhaust fumes that have been drawn back into the building. Worn or poorly maintained combustion devices (e.g., boilers, furnaces), or a flue that is improperly sized, blocked, disconnected, or leaking, can be significant sources. Auto, truck, or bus exhaust from attached garages, nearby roads, or idling vehicles in parking areas can also be sources.

Exposure to concentrated levels of CO may result in a variety of flu-like symptoms such as dizziness, fatigue, headaches, disorientation and nausea. High levels of exposure can result in loss of consciousness and death.

Combustion equipment must be maintained to assure that there are no blockages, and air and fuel mixtures must be properly adjusted to ensure more complete combustion. Vehicular use should be carefully managed adjacent to buildings and in vocational programs. Additional ventilation can be used as a temporary measure when high levels of CO are expected for short periods of time.

Steps to Prevent Carbon Monoxide Exposures:

✔ Annually inventory and inspect all gas burning appliances such as stoves, furnaces and water heaters to ensure they are properly operating and vented to the outside.

✔ Install carbon monoxide alarms in the school near appliances that burn natural gas, oil, wood or gas.

✔ Never let school buses or other vehicles idle directly outside of the school, particularly in places where air can get indoors such as air handling intakes, windows or exit doors.

Learn more at:

www.epa.gov/iaq/schools/tfs/guidee.html#Carbon%20 Monoxide

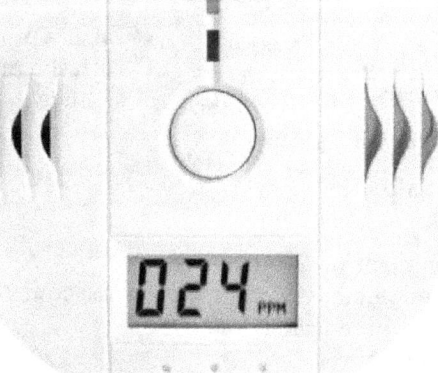

Chemical Management

From elementary school maintenance closets to high school chemistry labs, schools use a variety of chemicals. When they are mismanaged, chemicals can put students and school personnel at risk from spills, fires, and other accidental exposures. Common hazardous chemicals include corrosive acids, bases, oxidizers, compressed gases and flammable solvents.

Chemical accidents impact children's safety, can cost thousands of dollars to clean up, disrupt school schedules and could even temporarily close schools. Toxic chemicals can cause serious health effects, including cancer; brain and nervous system disorders; organ damage (i.e., liver, kidneys, and lungs); irritation of the eyes, skin nose and throat; and asthma attacks.

A proper chemical management program ensures that all schools are free from hazards associated with mismanaged chemicals. Chemicals may be considered mismanaged when they are:

- In poor condition or expired
- Overabundant
- Not needed or used
- Not properly labeled or unknown
- Unsecured
- Stored near food
- Stored in inappropriate, leaking, corroded or cracked containers
- Stored with incompatible chemicals
- Stored on unstable/incompatible shelves or cabinets

Responsible chemical management programs start with development and implementation of a safe chemical management plan that reduces the risk of chemical exposures and accidents in schools. Proper chemical management includes: a strong inventory control process, assessment of chemicals for risk and benefit, prohibiting the use of unauthorized chemicals, proper hazardous chemicals cleanout and disposal, appropriate safety measures, personal protective equipment and eyewash/emergency shower stations. Every chemical used in a school must have an accompanying Material Safety Data Sheet (MSDS) that describes chemical properties, potential hazards, storage, disposal, protective equipment and spill handling procedures.

Steps to Improve Chemical Management Include:

✔ Conduct annual chemical inventories and prohibit any unauthorized, toxic or hazardous chemicals from being brought into the school.

✔ Store toxic or hazardous chemicals in appropriate containers, separated by hazard category in a ventilated, fire resistant, and locked area or cabinet.

✔ Label containers with the name of the material, date it entered the school, and ensure that an MSDS for each product is in a binder readily displayed near the chemical storage area.

✔ Conduct regular cleanouts of chemicals that are unnecessary, outdated, and pose a health, safety or environmental risk.

✔ Ensure proper training of staff involved with chemical management and training of students before using toxic or hazardous chemicals.

Learn more at:

www.epa.gov/wastes/partnerships/sc3/

Drinking Water

Clean drinking water is necessary for good health. Harmful chemicals and micro-organisms in school drinking water can pose a threat to the health of students and staff. Although the majority of schools receive drinking water from public water supplies, EPA estimates that approximately 10,000 schools and child care facilities maintain their own water supply and are regulated under the Safe Drinking Water Act (SDWA).

While the vast majority of public drinking water systems are safe and dependable, drinking water pipes, taps, solder and other plumbing components may contain lead. Lead in plumbing components may leach into water and pose a health risk when consumed. Some drinking fountains have been shown to have high levels of lead leaching from their interior components.

Exposure to lead is a significant health concern, especially for young children and infants whose growing bodies absorb more lead than the average adult. Testing water in schools and child care facilities is important because children will likely be drinking water in school.

Water from public water supply systems is regularly tested to ensure it meets federal and state drinking water standards. School administrators of on-site well water systems are responsible for making sure the water is safe. This includes protecting the source from contamination, regularly testing and reporting monitoring results, and maintaining the distribution system.

Steps to Ensure Safe Drinking Water:

✔ Comply with state and federal drinking water standards if your school receives its drinking water from your own water source. Determine your sampling requirements and test your water as required.

✔ Never dispose of hazardous substances by flushing them down toilets or dumping them into storm drains.

✔ Maintain and sanitize water fountains and faucet screens/aerators regularly.

Learn more at:

http://water.epa.gov/infrastructure/drinkingwater/schools/index.cfm

http://epa.gov/safewater/lead

Educational, Art and Science Supplies

Common K-12 classroom supplies and those used in art, science labs, and vocational/technical education instruction, are an important part of the educational process. These supplies could include glues, cleaners, glazes, paints, solvents, and other materials helpful to classroom instruction. Many of these materials are formulated with hazardous ingredients which can be harmful or toxic to children when used improperly or by an inappropriate age group.

Dangerous metals such as lead, volatile organic compounds, dust and fibers are commonly found in the art materials and supplies in ordinary classrooms. Ingestion and skin absorption can occur when handling these materials as well as many other hazardous products. In fact, it is not uncommon for students to consciously "sniff" and smell the odors associated with contact cement, glues, paint thinners, correction fluids and solvents.

Simple preventive measures can greatly reduce harmful exposures to students. Careful purchase and selection of art materials, dedicated adult supervision, and the proper use of the product with the appropriate age group are all simple actions that should be implemented. For added security, always lock up chemicals when they are not in use. The U.S. Consumer Product Safety Commission recommends that when buying art materials, school supplies and toys you should only purchase those products that are labeled "Conforms to the American Society for Testing and Materials (ASTM) D4236."

Steps to Reduce Exposure to Hazards Associated With Educational, Art and Science Supplies:

✔ Check whether your supplies are listed as toxic or nontoxic (should be labeled accordingly by the Art and Creative Materials Institute (ACMI)).

✔ Read and follow directions on labels regarding fumes or ventilation.

✔ Ensure you have read and have available the Material Safety Data Sheets (MSDS) for all products being used in the classroom.

✔ Provide ample fresh air and ventilation.

✔ Do not allow eating or drinking around hazardous chemicals.

✔ Properly store and dispose of all products according to label instructions.

✔ Wash hands often when using toxic or hazardous chemicals.

Learn more at:

www.epa.gov/iaq/schools/pdfs/kit/checklists/teacherchklstbkgd.pdf

Extreme Heat Events

Extreme heat events, or heat waves, are defined by weather that is substantially hotter and/or more humid than average for a location at that time of year. These conditions stress the body's ability to maintain an ideal internal temperature, which can lead to a range of adverse health effects. School districts should consider starting the school year later in the season to avoid the heat of the summer.

Children need to take extra precautions on days of extreme heat. Dehydration, heat stroke, and other heat illnesses may affect a child more severely than the average adult because:

- Children have a smaller body mass to surface area ratio than adults, making them more vulnerable to the heat.

- Children are more likely to become dehydrated than adults because they can lose fluid more quickly.

- Children play outside more than adults, and they may be at greater risk of heat stroke and exhaustion because they may lack the judgment to limit exertion during hot weather and to rehydrate themselves.

Hot weather can also affect ozone levels and other types of air quality. The Air Quality Index (AQI) is a guide for reporting daily air quality. The EPA Flag Program uses colored flags based on the AQI to teach coaches, students and others about outdoor air quality conditions. Schools raise a colored flag each day that corresponds to their local air quality forecast. To check for air quality conditions in your area, go to: **http://airnow.gov**.

Steps to Protect Children from Extreme Heat:

✔ Limit outdoor activity and organized athletic events to morning hours when possible.

✔ Encourage students to drink more fluids.

✔ Advise students to wear lightweight, light-colored, loose-fitting clothing.

✔ Limit physical exertion outdoors during days with high heat or unhealthy air conditions.

✔ Plant more trees and vegetation (low pollination varieties) on school grounds.

✔ Participate in the EPA's colored flag program to clearly communicate the daily AQI to students, staff and others.

Learn more at:

www.epa.gov/heatisland/about/heatguidebook.html

Indoor Air Quality/ Ventilation

Poor indoor air quality (IAQ) can impact the comfort and health of students and staff which can affect concentration, attendance and student performance. Additionally, if schools fail to respond promptly to poor IAQ, students and staff are at an increased risk of short-term health problems.

Inadequate IAQ can result in health concerns such as fatigue, nausea, coughing, eye irritation, headaches, asthma episodes, allergic reactions, and in rare cases, life threatening conditions such as severe asthma attacks. Many perceived IAQ problems, however, are often comfort problems, such as temperature, humidity or air movement in a space being too low or too high.

Proper ventilation with outdoor air is a key component for good indoor air quality in schools and classrooms. In many cases, indoor air may, potentially, be two to five times more polluted than outdoor air. While at times challenging due to the high occupant densities of schools and classrooms, it is important that building designers incorporate ventilation systems that provide adequate outdoor ventilation air complying with the American Society of Heating, Refrigerating and Air Conditioning Engineers' standard, (ASHRAE) 62.1-2010 or local codes.

Factors that contribute to poor IAQ in schools may originate from inadequate heating, ventilation and air conditioning (HVAC) design. Some may be solely in the control of facilities management, such as proper maintenance of the HVAC system and the amount of outside air being mechanically brought into the building. The cleanliness and general housekeeping of a school building is a shared responsibility and requires the cooperation of facility management as well as the staff who work in the building. Many of the topics discussed in this brochure, such as pesticides, idling, and chemical management are all factors that influence a facility's IAQ. The goal of an Indoor Air Quality Management Program is to prevent the occurrence of IAQ problems and to respond quickly to issues before they become serious health matters.

Steps to Improve IAQ and Ventilation:

✔ Please ensure the school ventilation system is operating as designed.

✔ Implement a proactive IAQ management program such as the Indoor Air Quality (IAQ) Tools for Schools program highlighted in website below.

✔ Develop and implement a tobacco-free campus policy.

✔ Establish and implement a regular schedule for maintaining unit ventilators, replacing air filters, cleaning supply air diffusers, return registers and outside air intakes and commission the HVAC system a minimum of once every 5 years.

✔ Ensure condensate pans are clean, unobstructed, and drain properly.

✔ Keep unit ventilators clear of books, papers and other items that can obstruct air flow.

Learn more at:

www.epa.gov/iaq/schools/actionkit.html

www.epa.gov/iaq/schools/facilities.html

Lead

Lead is a highly toxic metal that can have adverse health effects for both children and adults. The most common source of lead is from paint in buildings built before 1978. Lead dust comes from disturbing lead paint during renovations, deteriorating lead paint and lead-contaminated soil that gets tracked into a school.

Children under 6 years of age are at particular risk of lead poisoning because their bodies are still developing. Furthermore, they frequently place their hands, toys, and other objects that could have dust from lead paint in their mouths. Some playground equipment and toys may contain lead or lead paint. Toys can pick up lead from contaminated soil or dust. Exposure to lead can result in lower IQ scores in children and has been associated with headaches, slowed growth, hearing problems, brain damage, nervous system disorders and behavior and attention problems.

If a school building was built before 1978, there is a good chance that it contains lead paint. EPA's Renovation, Repair and Painting Rule (RRP) is directed to child occupied facilities (COF) built before 1978 and applies to buildings where there are children under 6 years of age. Get your pre-1978 COFs tested for lead paint by a certified inspector or risk assessor. The RRP Rule applies only in areas that have not been tested for lead paint or are shown to contain lead paint after testing.

RRP Rule Requirements:

- Renovators must provide building owners and occupants pre-renovation notification.

- Firms/contractors and school staff performing renovations that disturb paint must be appropriately certified.

- Renovators must be trained and certified.

- Workers must receive on-the-job training from a certified Renovator.

- Lead-safe work practices must be followed and documented.

Steps to Reduce Lead Exposures:

✔ Interior painted areas—Examine walls and interior surfaces to see if the paint is cracking, chipping, or peeling, and check for areas on doors or windows where painted surfaces rub together.

✔ Exterior painted areas—Check exterior paint for flaking and ensure it is not contaminating nearby soil where children may play.

✔ Check large outdoor structures for peeling or flaking paint that could contaminate the soil around play areas.

✔ Have staff ensure that children wash their hands thoroughly after playing outside and before eating.

Learn more at:

www.epa.gov/lead
www.LeadFreeKids.org

Mercury

Mercury is commonly found in schools. Elemental mercury is found in thermometers, barometers, switches, thermostats, and glass vials. Mercury salts are found in laboratory compounds in chemistry and science laboratories. Compact Fluorescent Lamps (CFLs) also contain mercury.

Mercury spills at schools are often caused by improper storage and mishandling of these items. Mercury is more likely than other lab chemicals to be misused, spilled and spread throughout schools. These types of exposures can occur when elemental mercury is spilled or when products that contain elemental mercury break and expose mercury to the air, particularly in warm or poorly-ventilated indoor spaces.

Mercury is a neurotoxic substance that can produce a wide range of health effects in children depending on the amount and timing of exposure. Elemental (metallic) mercury primarily causes health effects when it is inhaled as a vapor and absorbed into the lungs.

Cleaning up mercury spills in schools can be costly and cause widespread environmental contamination since it can easily be tracked through a building and to other buildings, vehicles, and personal property (e.g. clothes, backpacks, toys). Whenever possible, items containing elemental mercury should be replaced in schools with alternatives such as digital thermometers.

Steps to Prevent Mercury Exposure:

✔ Conduct an inventory of all chemicals and locate all mercury equipment and compounds.

✔ Contact a professional to collect and properly dispose of all mercury equipment and compounds.

✔ In the event of a spill, have everyone leave the area, open windows, turn down the temperature and contact local or state health or environmental agencies and go to: www.epa.gov/hg/spills/index.htm

✔ Create and distribute a mercury spill response plan. www.epa.gov/region7/mercury/educator_toolkit.htm

✔ Spills—the size of a single thermometer or CFL—can be cleaned by school personnel after opening a window and ventilating the area. To learn proper cleanup and disposal procedures, go to www.epa.gov/hg/spills/index.htm#thermometer and www.epa.gov/cfl/cflcleanup.html

✔ Never use a vacuum cleaner to clean up mercury. The vacuum will put more mercury into the air and increase exposure.

✔ Never use a broom to clean up mercury. It will break the mercury into smaller droplets, spread them, and contaminate the broom.

✔ Never wash clothing or other items that have come in direct contact with mercury in a washing machine, because mercury may contaminate the machine and/or pollute the sewage system. Clothing that has come into direct contact with mercury should be discarded as directed by your local health or fire department.

Learn more at:
www.epa.gov/hg/schools.htm
www.epa.gov/mercury

Mold and Moisture Control

Individual school districts have incurred costs from $200,000 to as much as $13 million for remediating mold and mildew damage. Potential health concerns are also an important reason to prevent mold growth and to clean up existing indoor mold growth.

All molds have the potential to cause health effects that may include irritation of the eyes, skin, nose, throat, and lungs of both mold allergic and non-allergic people. Molds can produce allergens that trigger allergic reactions or even asthma attacks in people allergic to mold. Others are known to produce potent toxins and/or irritants. Molds can be found almost anywhere; and they can grow on virtually any organic substance, as long as moisture and oxygen are present. There are molds that can grow on wood, paper, carpet, foods and insulation.

The presence of moisture within school and building structures stimulates the growth of molds and other biological contaminants. The key to mold control is moisture control. Moisture and uncontrolled humidity problems may include roof leaks, landscaping or gutters that direct water into or under the school building, and unvented combustion appliances. Additionally, moist school facilities provide a nurturing environment for mites, roaches and rodents which are associated with asthma and other diseases. Solve moisture and condensation problems before they become mold problems.

Steps to Prevent Mold and Control Moisture:

✔ Maintain indoor humidity levels below 60%, ideally between 30% and 50% when possible.

✔ Clean and dry any wet or damp spots within 48 hours.

✔ Fix leaky plumbing and roof leaks in the school as soon as possible.

✔ Check regularly for condensation and wet spots.

✔ Address sources of moisture problems as soon as possible.

✔ Scrub mold off hard surfaces with water and detergent, and dry completely.

Learn more at:

www.epa.gov/mold

www.epa.gov/iaq/schools/pdfs/publications/moldfactsheet.pdf

PCBs in Caulk and Fluorescent Light Ballasts

Polychlorinated biphenyls (PCBs) are a class of organic chemicals that have been used in a variety of commercial products. PCBs were used in caulking, electronics, fluorescent light ballasts and other building materials from the 1950s to the late 1970s. Buildings built or renovated during that time may contain PCBs in caulking and other materials.

In 1979, the U.S. Environmental Protection Agency (EPA) banned the commercial production of PCBs, citing health and environmental concerns. EPA has found that PCB-containing caulk and PCB containing lighting ballasts can be a significant source of PCBs in school air. Health concerns related to PCB exposure include, but are not limited to, cancer, reproductive effects and neurological effects.

Caulk is a flexible material used to seal gaps and to make airtight or watertight windows, door frames, masonry and joints in buildings and other structures. EPA found old caulk in schools that were constructed or renovated between 1950 and 1979 may contain as much as 30% PCBs and can emit PCBs into the surrounding air. PCBs from caulk may also contaminate adjacent materials such as masonry or wood.

PCBs are also contained within some fluorescent light ballast capacitors and potting material manufactured prior to 1979. PCB-containing fluorescent light ballasts that are currently in use have either approached or exceeded their designed life span, so they should be properly removed from buildings to prevent indoor air exposure. Sudden rupture of light ballasts may pose health risks to the occupants, and is difficult and costly to remediate. Removal of PCB-containing light fixtures, as part of lighting upgrades or a stand-alone project, is an investment that pays off with long-term benefits to students, school staff, the community, and the environment.

Steps to Minimize Exposure to PCBs:

✔ Clean frequently, using a damp cloth or mop to reduce dust.

✔ Use HEPA vacuums.

✔ Wash children's hands with soap and water, particularly before eating, and wash their toys often.

✔ Consider getting a professional to test the air and if elevated levels of PCBs are found, schools should identify any potential sources of PCBs, for example by testing samples of caulk or looking for other potential PCB sources (i.e. old transformers, capacitors, or fluorescent light ballasts).

What NOT to Do:

✔ Do not attempt to remove PCB-containing caulk or lighting ballasts by yourself. PCBs should be removed by personnel wearing protective equipment who should follow procedures to minimize the spread of PCBs.

✔ Do not sweep with dry brooms; minimize the use of dusters because they spread dust.

Learn more at:

http://epa.gov/pcbsincaulk/caulkschoolkit.htm

www.epa.gov/epawaste/hazard/tsd/pcbs/pubs/ballasts.htm

Pesticides and Pest Management

Pesticides need to be used carefully and judiciously, especially when used in sensitive areas where children are present. Children are more sensitive than adults to pesticides. In addition, young children can have greater exposure to pesticides from crawling, exploring, or other hand-to-mouth activities.

Adverse effects of pesticide exposure range from mild symptoms of dizziness and nausea to serious, long-term neurological, developmental and reproductive disorders.

EPA recommends that schools use an Integrated Pest Management (IPM) approach to reduce pesticide risk and exposure to children and staff. Implementing IPM practices in schools can reduce or minimize economic and health related issues caused by pests and pesticides.

All school occupants and employees play a role in ensuring that a school's IPM program is successful. Ask school administrators if an established IPM program is being utilized in your school. By working together, everyone can have a role in creating an on-going safe and healthy school environment.

Learn more at:

www.epa.gov/pesticides/ipm/schoolipm

Steps to Reduce Pesticides and Manage Pests in your school:

The central features of an IPM program are the implementation of exclusion and sanitation practices that keep pests out.

Exclusion Practices:

✔ Install high-density door sweeps on all doors to keep mice, rats and roaches out.

✔ Block open spaces around utility pipes coming into the building with copper mesh wire. Open spaces as small as ¼ inch, or about the height of a dime, will allow mice and other pests into a building.

✔ Install screens on all windows, particularly if they are open during warm months.

Sanitation Practices:

✔ Clean and mop floors in all food service areas daily, including classrooms.

✔ Use sealable containers or canisters to provide secure storage for edible food items and snacks.

✔ Bag and completely close all garbage and place in dumpsters outside of the school building daily.

✔ Bag and completely close all garbage and place in dumpsters outside of the school building daily.

Radon

Radon is a radioactive, colorless, and odorless gas that comes from the natural (radioactive) breakdown of uranium in soil, rock and water. Radon gas can enter a building through cracks and holes in the floor and become trapped in indoor air. Radon can be found in both old and new buildings and cannot be felt when inhaled into your lungs.

Prolonged exposure to radon can result in lung cancer. Higher radiation doses may result in children due to their smaller bodies and faster breathing rates compared to adults. The EPA estimates that radon is responsible for 21,000 lung cancer deaths every year making radon the second leading cause of lung cancer in the U.S., after smoking.

Radon test kits cost $10-15 and can be purchased from environmental laboratories, local hardware stores and building supply companies or through the National Radon Hotline, (800) 767-7236.

Learn more at:

www.epa.gov/radon/index.html
www.sosradon.org

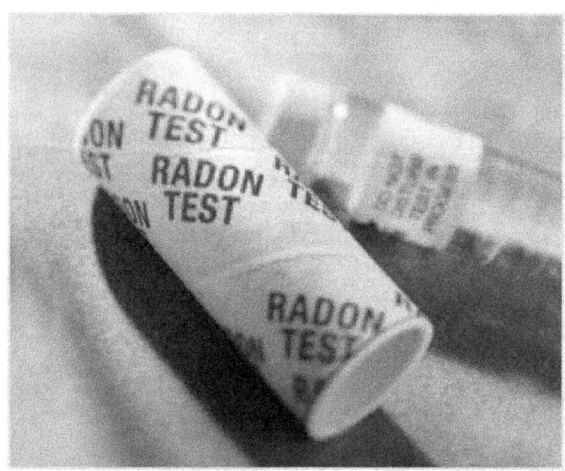

Steps to Reduce Radon Exposure:

✔ Have authorized personnel test classrooms and occupied rooms for radon, following EPA and State protocols.

✔ Install radon mitigation systems which, are designed to reduce and remove radon from indoor air if the classrooms testing results show radon concentrations of 4 pCi/L or higher. For information on reducing radon levels in schools go to: www.epa.gov/radon/pubs/#schools

UV Radiation

While short exposure to sunlight is enjoyable and beneficial as an important source of Vitamin D, too much exposure to the sun can be dangerous. Most people are not aware that skin cancer, while largely preventable, is the most common form of cancer in the United States. In fact, more than 3.5 million new cases of skin cancer are reported each year.

Overexposure to ultraviolet (UV) radiation from the sun can result in painful sunburns. It can also lead to more serious health problems, including skin cancer, premature aging of the skin, cataracts and other eye damage, and immune system suppression. Because they tend to play outside more frequently for long periods of time and may not have the benefit of sunscreen or shade, children are particularly at risk. By following some simple steps, children and staff can still enjoy time in the sun and be protected from overexposure to UV radiation.

Learn more at:

www.epa.gov/sunwise

Steps to Protect Students and Staff from Overexposure to UV Radiation:

✔ Take steps to prevent sunburns. Sunburns significantly increase a person's lifetime risk of developing skin cancer, especially for children.

✔ Wear protective clothing. A long-sleeved shirt, a wide brimmed hat, and sunglasses are strongly recommended.

✔ Generously apply broad-spectrum sunscreen with a minimum Sun Protection Factor (SPF) of 30+ approximately 15 minutes prior to going outside. Sunscreen should provide protection from both ultraviolet A (UVA) and ultraviolet B (UVB) rays. Reapply every two hours, even on cloudy days, and after swimming or sweating.

✔ Designated school personnel should check the UV index which is issued daily by EPA and the National Weather Service. This index provides important information to help plan for outdoor activities in ways that prevent sun overexposure.

✔ Provide access to shade on school grounds and remember that the sun's UV rays are strongest between 10 a.m. and 4 p.m.

UV INDEX	
Exposure Category	**UV Index Range**
Low	2 or less
Moderate	3 to 5
High	6 to 8
Very High	8 to 10
Extreme	11+

Energy Efficiency

The nation's 17,450 K-12 school districts spend more than $8 billion annually on energy — more than is spent on computers and textbooks combined. As much as 30 percent of a district's total energy is used inefficiently or unnecessarily.

By being more energy efficient, schools can save money and prevent greenhouse gas emissions. School districts can use the savings from improved energy performance to help pay for building improvements and other upgrades that enhance the learning environment.

Learn more at:

www.energystar.gov/k-12

Energy Efficiency Opportunities for Schools:

Low-Cost Measures:

✔ Use EPA's measurement and tracking tool, Portfolio Manager, to assess energy performance. Go to: **www.energystar.gov/index. cfm?=evaluate_performance.bus_portfoliomanager**

✔ Turn off lights when not in use or when natural daylight can be used.

✔ Set back the thermostat in the evening and at other times when the building is unoccupied.

✔ Perform monthly maintenance of heating and cooling equipment to ensure efficient operation throughout the year.

✔ Educate students and staff about how their behaviors affect energy use. Some schools have created student energy patrols to monitor and inform others when energy is wasted.

✔ Use Energy Star's Commercial Building Design Resource, Target Finder, to set energy targets and integrate efficiency goals into the design of new properties. Go to: **www.energystar.gov/index.cfm?c=new_bldg_ design.bus_target_finder**

Cost-Effective Investments:

✔ Install energy-efficient lighting systems and controls which will improve light quality, and reduce heat gain. Installing new energy-efficient lighting systems will also serve to remove any potentially harmful PCB-containing light ballasts.

✔ Upgrade and maintain heating and cooling equipment.

✔ Use a performance-based contract to guarantee energy savings from upgrades.

✔ Work with an energy services provider to help manage and improve energy performance.

✔ Purchase energy-efficient products like ENERGY STAR qualified office equipment.

✔ Install window films and add insulation or reflective roof coating.

Waste Reduction

Americans generate millions of tons of trash in our homes and communities. Every day, the average individual living in the United States produces approximately 4.5 pounds of trash. While many people already recycle products at home, schools can also control their waste by reducing, reusing and recycling it. Waste reduction opportunities exist everywhere.

Products that can be reused and recycled are countless and include everything from food scraps, yard and grounds wastes, paper, clothing, school supplies, sports equipment, and electronics. Items commonly recycled, and found in virtually any school, are paper, aluminum, glass, steel, cardboard, and yard waste. Food scraps or yard waste can be composted instead of being thrown out and then be used to improve the soil and support school landscaping or gardens. Many waste reduction efforts save money, energy, and natural resources, while teaching children and young adults how solid waste affects their lives and their environment.

Schools have a tremendous opportunity to implement waste reduction programs that can include pre-post waste reduction audits, incorporate waste tracking activities into the science curriculum, and promote programs that reduce waste. Engage school-related groups including science classes, environmental clubs, and parent-teacher organizations. These groups can often educate the whole community about the benefits of waste reduction and encourage everyone to make waste reduction a part of their everyday life. Increasing the flow of reusable and recyclable materials can even generate extra funds for school departments and groups.

Waste reduction can be further minimized by using WasteWise, a free EPA program through which organizations eliminate costly municipal solid waste.

To address electronic waste, refer to the resources available from the State Electronics Challenge.

Steps to Reduce Waste:

✔ Reduce waste through recycling, reusing or buying recycled products. Visit EPA's What Can You Do webpage for simple tips on how to make a difference in the environment whether at home, school, work, or on the go. www.epa.gov/wastes/wycd/index.htm

✔ Purchase more environmentally friendly electronic and paper products.

✔ Manage obsolete electronics in an environmentally safe way.

✔ Purchase less food to save money and reduce waste.

Learn more at:

www.epa.gov/epawaste/conserve/smm/wastewise/index.htm

www.stateelectronicschallenge.net/

http://epa.gov/wastes/education/pdfs/school.pdf

Quick Assessment

Please use this voluntary assessment foldout to help you reduce and prevent exposures to common environmental health hazards in your school. Each topic area covered below has low or no-cost steps which can be taken to improve your students' environmental health. This tool also highlights waste reduction and energy efficiency strategies to help conserve valuable, financial resources.

Asbestos

Does the school have a readily available asbestos management plan?
○ Yes ○ No ○ N/A

Have all building operations and maintenance staff reviewed the asbestos management plan and understand how to minimize potential disturbance to ACM?
○ Yes ○ No ○ N/A

Asthma and Asthma Triggers

Are there policies that discourage the use of birds or furry animals such as mice, guinea pigs, and rabbits as pets in the school's classrooms?
○ Yes ○ No ○ N/A

Is the school's cleaning staff encouraged to use environmentally friendly cleaning products and "wet" dusting techniques whenever possible?
○ Yes ○ No ○ N/A

Are classrooms free of clutter? Are they dusted regularly? Are stuffed animals and pillows washed frequently?
○ Yes ○ No ○ N/A

Buses and Vehicle Idling

Have anti-idling policies been developed and implemented for buses that serve the school?
○ Yes ○ No ○ N/A

Have anti-idling zones been established for all vehicles at the schools (school buses, delivery trucks and parent's cars)?
○ Yes ○ No ○ N/A

Are all passenger pickup/drop off areas located away from school's air intake supply, classroom windows and exit doors?
○ Yes ○ No ○ N/A

Carbon Monoxide (CO)

Does the school's maintenance staff inspect and document the condition and findings for all gas burning appliances, furnaces and water heaters yearly to ensure they are properly operating?
○ Yes ○ No ○ N/A

Have CO detectors been installed in the school near appliances that burn natural gas, oil, wood or gas?
○ Yes ○ No ○ N/A

Chemical Management

Does the school conduct a yearly inventory of all chemicals present?
○ Yes ○ No ○ N/A

Does the school have a policy that prohibits any unauthorized toxic or hazardous chemicals from being brought into the school?
○ Yes ○ No ○ N/A

Are all chemicals properly labeled, stored in original containers, dated as to when they entered the school, and have accompanying MSDS information on site?
○ Yes ○ No ○ N/A

Are all toxic or hazardous chemicals stored in appropriate containers, separated by hazard category, in a ventilated, fire resistant, and locked area or cabinet?
○ Yes ○ No ○ N/A

Does the school conduct cleanouts of all chemicals that are unnecessary, outdated and posing a health risk on a regular basis?
○ Yes ○ No ○ N/A

Does the school ensure proper training of staff involved with chemical management? Are students properly trained before handling toxic or hazardous chemicals?
○ Yes ○ No ○ N/A

Drinking Water

If your school receives its drinking water from your own source, you are required to comply with a series of regulations under the Safe Drinking Water Act. Is the water tested according to regulations and the results documented?
○ Yes ○ No ○ N/A

Are water faucets, fountain screens and aerators regularly cleaned and sanitized?
○ Yes ○ No ○ N/A

Does the school have policies and procedures in place to prevent the disposal of hazardous substances down the toilets and/or dumping into storm drains?
○ Yes ○ No ○ N/A

Educational, Art & Science Supplies

Does your school have a policy to ensure that art materials, school supplies and toys purchased are labeled "Conforms to ASTM D4236"?
○ **Yes** ○ **No** ○ **N/A**

Do school policies encourage minimizing exposure to hazardous materials by substituting less- or nonhazardous materials where possible for classroom activities; prohibiting food consumption around hazardous chemicals; and, washing hands often?
○ **Yes** ○ **No** ○ **N/A**

Is the school staff reminded to follow the precautionary recommendations listed on the labels?
○ **Yes** ○ **No** ○ **N/A**

Does the school have updated Material Safety Data Sheets for all products being used?
○ **Yes** ○ **No** ○ **N/A**

Energy Efficiency

Are lights turned off when not in use or when natural daylight can be used?
○ **Yes** ○ **No** ○ **N/A**

Are thermostats set back in the evening and at other times when the building is unoccupied?
○ **Yes** ○ **No** ○ **N/A**

Does the school track energy performance, perform monthly maintenance of heating and cooling equipment, educate students and staff about how their behaviors affect energy use, and use systems and controls that improve light quality, heating and cooling as part of an energy efficiency program?*
○ **Yes** ○ **No** ○ **N/A**

Extreme Heat Events

Does the school advise the students to wear lightweight, light-colored and loose-fitting clothing during extreme heat events?
○ **Yes** ○ **No** ○ **N/A**

Does the school limit physical exertion outdoors during days with unhealthy air conditions or periods of extreme heat?
○ **Yes** ○ **No** ○ **N/A**

Indoor Air Quality/Ventilation

Does the school currently implement a proactive IAQ management program such as the "Indoor Air Quality (IAQ) Tools for Schools" program?
○ **Yes** ○ **No** ○ **N/A**

Does the school have a tobacco-free campus policy?
○ **Yes** ○ **No** ○ **N/A**

Does the school maintenance staff have a regular cleaning schedule for unit ventilators, supply air diffusers, return registers, outside air intakes, and commission the HVAC system a minimum of once every 5 years?
○ **Yes** ○ **No** ○ **N/A**

Are condensate pans clean, unobstructed and do they drain properly?
○ **Yes** ○ **No** ○ **N/A**

Are unit ventilators clear of books, papers and other items and other items that would block or hinder air flow?
○ **Yes** ○ **No** ○ **N/A**

Lead

Are the walls and interior surfaces free of cracking, chipping, or peeling paint, especially around doors or windows where painted surfaces rub together?
○ **Yes** ○ **No** ○ **N/A**

Are exterior walls and other large structures in the school grounds free of cracking, chipping, or peeling paint?
○ **Yes** ○ **No** ○ **N/A**

Does the school or school district provide a pre-renovation notification to staff and parents prior to construction activity?
○ **Yes** ○ **No** ○ **N/A**

Are all demolition and renovation activities impacting lead containing paint or other building materials in the school undertaken by "certified" and properly trained contractors?*
○ **Yes** ○ **No** ○ **N/A**

Do students wash hands before snacks, lunch and after recess?
○ **Yes** ○ **No** ○ **N/A**

Mercury

Has an inventory of all chemicals, materials and equipment containing mercury been completed?
○ **Yes** ○ **No** ○ **N/A**

Does the school have a mercury spill kit and spill response plan readily available on site?
○ **Yes** ○ **No** ○ **N/A**

Mold and Moisture Control

Is humidity in the school building maintained below 60%, and ideally between 30% and 50% where possible?*
○ **Yes** ○ **No** ○ **N/A**

Does the school maintenance staff repair all leaking plumbing and roof leaks in the building as soon as possible?*
○ **Yes** ○ **No** ○ **N/A**

Is the school building (walls/ceilings/floors) free of wetness or condensation?
○ **Yes** ○ **No** ○ **N/A**

Does the school maintenance staff clean and dry any wet or damp spots consistently within 48 hours?
○ Yes ○ No ○ N/A

PCBs in Caulk and Fluorescent Light Ballasts

Has the school determined whether the fluorescent light ballasts contain PCBs? If so, have the lighting fixtures in the school been retrofitted to adequately remove potential PCB hazards using recommendations highlighted in **www.epa.gov/epawaste/hazard/tsd/pcbs/pubs/ballasts.htm**?
○ Yes ○ No ○ N/A

Has the school followed recommendations highlighted in **www.epa.gov/pcbsincaulk/caulkschoolkit.htm** for potential PCB-containing caulk.
○ Yes ○ No ○ N/A

Pesticides and Pest Management

Do all floors in food service areas and classrooms where food is served get cleaned and mopped daily?
○ Yes ○ No ○ N/A

Are all food items stored securely in sealable containers or canisters?
○ Yes ○ No ○ N/A

Is all garbage bagged, completely closed, and placed in dumpsters outside the school building daily?
○ Yes ○ No ○ N/A

Are there high-density door sweeps installed on all doors to keep out mice, rats and roaches?*
○ Yes ○ No ○ N/A

Are all open spaces around utility pipes coming into the building blocked with copper mesh wire or other material to hinder entrance into the building by pests?*
○ Yes ○ No ○ N/A

Does the school have screens installed on all operable windows?*
○ Yes ○ No ○ N/A

Radon

Have all the first floor and basement classrooms of the school been tested for the presence of radon with results documented and available for public review?
○ Yes ○ No ○ N/A

If the classroom radon levels exceed 4pCi/L, has the school or school district installed radon mitigation systems?*
○ Yes ○ No ○ N/A

UV Radiation

Does the school post the daily UV Index to help staff protect student's overexposure to the sun?
○ Yes ○ No ○ N/A

Are students encouraged to wear protective, light weight clothing and/or sunscreen for recess during times of peak sun intensity?
○ Yes ○ No ○ N/A

Does the school have ample areas of shade to minimize time spent in the sunlight?
○ Yes ○ No ○ N/A

Waste Reduction

Does the school use an active waste reduction/recycling policy in place that promotes resource conservation, the purchasing of more environmentally friendly products, pre-post waste reduction audits, student involvement, and a curriculum that supports waste reduction and recycling?*
○ Yes ○ No ○ N/A

Assessment Activities that may require additional, cost-effective resources and methods for resolution.

Healthy School Environments Assessment Tool (HealthySEAT)

EPA has developed a free software tool to help school districts track and manage all environmental issues in their schools. The Healthy School Environments Assessment Tool (HealthySEATv2) is designed to be customized and used by district-level staff to conduct voluntary self-assessments of their school (and other) facilities, using checklists and assessment standards included with the software or customizing the tool to create their own checklists. The software generates letters and reports that can be used to follow up on any deficiencies found and to track progress in creating and maintaining healthy school environments. To learn more about HealthySEAT and how it can help improve the health and safety of students and staff, visit **www.epa.gov/schools/healthyseat/basicinformation.html**

Top Ten Ways to Make Your School Healthier

1. **Clear the air inside.** EPA's Indoor Air Quality Tools for Schools program provides information and tips on how to help schools prevent and solve indoor air quality problems.

2. **Clear the air outside.** Schools can reduce children's exposure to engine exhaust by eliminating unnecessary vehicle idling, installing effective emission control systems on newer buses and replacing the oldest buses with new ones.

3. **Reduce/remove radon in school buildings.** Schools should test the level of radon gas in their buildings. No radon level is healthy. If the test results are at, or above, 4pCi/L, appropriate mitigation steps should be taken to reduce the radon level.

4. **Use chemicals carefully.** Possible health, safety and environmental implications should be considered before chemicals are purchased for use in schools. Do not allow outside, unauthorized chemicals to be brought into the school. Proper chemical use and management (storage, labeling, and disposal) is critical for reducing chemical exposures and costly accidents.

5. **Test the water.** School districts should test the drinking water in their school buildings regularly.

6. **Get the lead out.** School buildings built prior to 1978 must be tested for lead paint. Renovations or repairs must be done in a way that does not create lead dust. Children should be kept away from lead hazards inside and outside of school buildings.

7. **Eliminate Mercury.** School environments should be mercury-free. Schools should use digital thermometers and safer alternatives to mercury in science curriculum, nurses' offices and within facilities operations and maintenance.

8. **Cover up.** Schools should practice "sunsafe behavior" and encourage children to cover up, use SPF 30 or higher broad-spectrum sunscreen, and stay out of midday sun to avoid damaging UV rays.

9. **Use toxics with caution.** Schools should look for alternatives to toxic pesticides and cleaning chemicals. Remove sources of lead, mercury, asbestos and PCBs from the school environment.

10. **Educate yourself.** Know which environmental health issues affect your school and how to address them.

Additional Online Resources

EPA Children's Environmental Health Website
Protecting children's health from environmental risks is fundamental to EPA's mission. Get the facts about children's environmental health at **www.epa.gov/children**

Pediatric Environmental Health Specialty Unit (PEHSU)
A respected network of experts in children's environmental health. The PEHSU were created to ensure that children and communities have access to, usually at no cost, special medical knowledge and resources for children faced with a health risk due to a natural or human-made environmental hazard. To learn more visit **www.pehsu.net**

Department of Education's Green Ribbon Schools
The U.S. Department of Education announced in 2011 the creation of the Green Ribbon Schools program to recognize schools that are creating healthy and sustainable learning environments and teaching environmental literacy. The new awards program will be run by the Education Department with the support of the White House Council on Environmental Quality and the U.S. Environmental Protection Agency. To learn more visit **www.ed.gov/blog/2011/05/green-ribbon-school-resources/**

EPA Regional School Contacts
Locate your regional EPA School Coordinator or Children's Health Coordinator by visiting **www.epa.gov/schools/contacts.html**

EPA's Voluntary School Siting Guidelines
EPA's voluntary school siting guidelines can help local school districts and community members evaluate environmental factors to make the best possible school siting decisions. This website includes an overview for the guidelines, as well as links to resources and additional information. **http://epa.gov/schools/siting/index.html**

EPA's Voluntary State School Environmental Health Guidelines
EPA has developed State School Environmental Health Guidelines, a voluntary guidance document which helps states, tribes, and territories create and implement environmental health programs for K-12 schools. The goal of the guidelines is to provide a framework for improving the health and well-being of students by creating and sustaining healthy, safe, and productive school environments. To learn more visit **www.epa.gov/schools/ehguidelines/index.html**

K-12 School Compliance
It is important to note that schools are obligated to comply with relevant environmental regulations, and environmental compliance is an integral part of a K-12 school environmental health program. To learn more visit **www.epa.gov/schools/downloads/key_USEPA_regulations_k-12_schools.pdf**